Biomass: Comparison of Definitions in Legislation Through the 112th Congress

Kelsi Bracmort
Specialist in Agricultural Conservation and Natural Resources Policy

November 14, 2012

Congressional Research Service

7-5700

www.crs.gov

R40529

CRS Report for Congress

Prepared for Members and Committees of Congress

Summary

The use of biomass as an energy feedstock is emerging as a potentially viable alternative to address U.S. energy security concerns, foreign oil dependence, rural economic development, and diminishing sources of conventional energy. Biomass (organic matter that can be converted into energy) may include food crops, crops for energy (e.g., switchgrass or prairie perennials), crop residues, wood waste and byproducts, and animal manure. Most legislation involving biomass has focused on encouraging the production of liquid fuels from corn. Efforts to promote the use of biomass for power generation have focused on wood, wood residues, and milling waste. Comparatively less emphasis has been placed on the use of non-corn-based biomass feedstocks— other food crops, non-food crops, crop residues, animal manure, and more—as renewable energy sources for liquid fuel use or for power generation. This is partly due to the variety, lack of availability, and dispersed location of non-corn-based biomass feedstock. The technology development status and costs to convert non-corn-based biomass into energy are also viewed by some as an obstacle to rapid technology deployment.

For over 30 years, the term *biomass* has been a part of legislation enacted by Congress for various programs, indicating some interest by the general public and policymakers in expanding its use. To aid understanding of why U.S. consumers, utility groups, refinery managers, and others have not fully adopted biomass as an energy resource, this report investigates the characterization of biomass in legislation. The definition of biomass has evolved over time, most notably since 2004. The report lists biomass definitions enacted by Congress in legislation and the tax code since 2004 and definitions contained in legislation from the 111ᵗʰ Congress (the American Clean Energy and Security Act of 2009, H.R. 2454; the American Clean Energy Leadership Act of 2009, S. 1462; the Clean Energy Jobs and American Power Act, S. 1733; and the discussion draft of the American Power Act). Comments on the similarities and differences among the definitions are provided. One point of contention regarding the definition is the inclusion of biomass from federal lands. Some argue that removal of biomass from these lands may lead to ecological harm. Others contend that biomass from federal lands can aid the production of renewable energy to meet certain mandates (e.g., the Renewable Fuel Standard) and that removal of biomass can enhance forest protection from wildfires. Factors that may prevent a private landowner from rapidly entering the biomass feedstock market are also included in the report.

Bills were introduced in the 112ᵗʰ Congress that would modify the biomass definition (e.g., S. 781, H.R. 1861). However, debates about the definition have not been as extensive in the 112ᵗʰ Congress as they were in previous Congresses. Forthcoming discussions about energy, particularly legislation involving the Renewable Fuel Standard or energy tax incentives, may prompt further discussion about the definition of biomass.

Contents

Tables

Contacts

Introduction

The potential for biomass to meet U.S. renewable energy demands has yet to be fully explored. Non-food and other types of biomass (e.g., manure) have traditionally been considered by some as waste material and as such have been deposited in landfills, used for animal feed, or applied to crop production lands. However, rising fuel prices, environmental concerns, and sustainability issues have led policymakers to create legislation that encourages conversion of biomass into liquid fuels (e.g., ethanol, biodiesel) or electricity.[1] Interest has increased in cellulosic biomass (e.g., crop residues, prairie grasses, and woody biomass) because it does not compete directly with crop production for food—although it may compete for land—and because it is located in widely dispersed areas.[2] Classification of biomass as an energy resource has prompted the investigation of its use for purposes additional to liquid fuel (e.g., on-site heating and lighting purposes, off-site electricity).

Biomass

Biomass is organic matter that can be converted into energy. Common examples of biomass include food crops, crops for energy (e.g., switchgrass or prairie perennials), crop residues (e.g., corn stover), wood waste and byproducts (both mill residues and traditionally noncommercial biomass in the woods), and animal manure. Over the last few years, the concept of biomass has grown to include such diverse sources as algae, construction debris, municipal solid waste, yard waste, and food waste. Some contend that biomass has seen limited use as an energy source thus far because it is not readily available as a year-round feedstock, is often located at dispersed sites, can be expensive to transport, lacks long-term performance data, requires costly technology to convert to energy, and might not meet quality specifications to reliably fuel electric generators.

Woody biomass has received special attention because of its widespread availability, but to date has been of limited use for energy production except for wood wastes at sawmills. Wood can be burned directly, usually to produce both heat or steam and electricity (called combined heat and power, or CHP), or digested to produce liquid fuels. Biomass from forests, as opposed to mill wastes, has been of particular interest, because it is widely accepted that many forests have excessive levels of biomass (compared to historic levels) called hazardous fuels that can contribute to catastrophic wildfires.[3] Removing these hazardous fuels from forests could reduce the threat of catastrophic wildfires, at least in some ecosystems, while providing a feedstock for energy production.

---

[1] For more information on biofuels and biopower, see CRS Report R40110, *Biofuels Incentives: A Summary of Federal Programs*, by Brent D. Yacobucci, CRS Report RL34239, *Biofuels Provisions in the 2007 Energy Bill and the 2008 Farm Bill: A Side-by-Side Comparison*, by Randy Schnepf and Brent D. Yacobucci; and CRS Report R41440, *Biomass Feedstocks for Biopower: Background and Selected Issues*, by Kelsi Bracmort.

[2] For more information on cellulosic biomass feedstocks, see CRS Report RL34738, *Cellulosic Biofuels: Analysis of Policy Issues for Congress*, by Kelsi Bracmort et al.; and CRS Report R41460, *Cellulosic Ethanol: Feedstocks, Conversion Technologies, Economics, and Policy Options*, by Randy Schnepf.

[3] See CRS Report R40811, *Wildfire Fuels and Fuel Reduction*, by Kelsi Bracmort.

Legislative History

The term *biomass* was first introduced by Congress in the Powerplant and Industrial Fuel Use Act of 1978 (P.L. 95-620) as a type of alternate fuel. Biomass was first defined in the Energy Security Act of 1980 (P.L. 96-294), in Title II, Biomass Energy and Alcohol Fuels, as "any organic matter which is available on a renewable basis, including agricultural crops and agricultural wastes and residues, wood and wood wastes and residues, animal wastes, municipal wastes, and aquatic plants." The Energy Security Act of 1980 contained two additional definitions for biomass, excluding aquatic plants and municipal wastes, in Title II, Subtitle C, Rural, Agricultural, and Forestry Biomass Energy.

Three germane pieces of recent legislation define biomass: the Food, Conservation, and Energy Act of 2008 (2008 farm bill, P.L. 110-246); the Energy Independence and Security Act of 2007 (EISA, P.L. 110-140); and the Energy Policy Act of 2005 (EPAct05, P.L. 109-58). The term is mentioned several times throughout the three bills, but is not always defined or referenced. In some cases, an individual law has multiple biomass definitions. For example, one definition is included in the 2008 farm bill and three are provided in EISA. EPAct05 has six biomass definitions. The tax code contains four definitions. A total of 14 biomass definitions have been included in legislation and the tax code since 2004. **Table 1** includes definitions from the three laws and from the tax code and contains additional comments.

The definitions are built into the many provisions and programs that may support research and development, encourage technology transfer, and reduce technology costs for landowners and businesses. Thus, because the various definitions determine which feedstocks can be used under the various programs, the definitions are critical to the research, development, and application of biomass used to produce energy.

Analysis of Biomass Definitions

Two biomass definitions may be considered by policymakers, scientists, and program managers as the most comprehensive for energy production purposes: the definition in Title IX of the 2008 farm bill and the definition in Title II of EISA. Both laws provide an extensive definition for renewable biomass, but each law defines renewable biomass somewhat differently. The recognition of biomass as renewable means that biomass is considered by some to be an infinite feedstock that may be replenished in a short time frame. Both definitions consider crops, crop residues, plants, algae, animal waste, food waste, and yard waste, among other items, as appropriate biomass feedstock.

An important distinction exists between the two definitions for renewable biomass. The 2008 farm bill includes biomass from federal lands as a biofuel feedstock. Under EISA, to be eligible for the Renewable Fuel Standard (RFS), biomass cannot be removed from federal lands, and the law excludes crops from forested lands.[4] There has been some congressional discussion and

---

[4] The Renewable Fuel Standard (RFS) is a provision established by the Energy Policy Act of 2005 requiring gasoline to contain a minimum amount of fuel produced from renewable biomass. For more information on the RFS, see CRS Report R40155, *Renewable Fuel Standard (RFS): Overview and Issues*, by Randy Schnepf and Brent D. Yacobucci.

legislation on expanding the EISA definition to include biomass from federal lands to better meet the biofuels usage mandate for the Renewable Fuel Standard.[5]

EISA expanded the RFS and restricted the definition of biomass. As described above, the renewable biomass definition for the RFS under EISA excludes biomass removed from federal lands and crops from forested lands as biofuel feedstocks. Advocates for this definition include groups who favor minimal land disturbance (for ecological reasons as well as to sustain sequestered carbon) and are concerned that incentives to use wood waste might increase land disturbance, especially timber harvesting on federal lands. Opponents of this definition include groups who seek to use materials from federal lands and other forested lands (i.e., not tree plantations) as a source of renewable energy while possibly contributing to long-term, sustainable management of those lands.

Advocates of the 2008 farm bill renewable biomass definition include groups who seek to use the potentially substantial volumes of waste woody biomass from federal lands and other (non-plantation) forest lands (e.g., waste from timber harvests, from pre-commercial thinnings, or from wildfire fuel reduction treatments) as a source of renewable energy. Opponents include groups who seek to preserve forested land and federal land, and who are concerned that incentives for using wood waste would encourage activities that could disturb forest lands, possibly damaging important wildlife habitats and water quality, as well as releasing carbon from forest soils.

Potential Issues for Biomass Feedstock Development

It is not clear whether the biomass definitions in the 2008 farm bill and in EISA constitute a barrier to biomass feedstock development for conversion to liquid fuels. Concerns for some landowners and business entities that wish to enter the biomass feedstock market include economic stability, risk/reward ratio, revenue generation, land use designation, and lifecycle greenhouse gas emissions. Additionally, the feedstock development potential of woody biomass varies by region. For example, biomass stock tends to be located on private forest land in the southeastern United States and on federal land in the western United States. To thrive, different regions may require different resources for the biomass feedstock market.

Recent agricultural and energy legislation has incorporated provisions and established programs to promote the development and use of biomass as a renewable energy source.[6] The success of these provisions and programs will be partly determined by landowner participation rates. Participation rates may depend on the definition provided in the legislation that authorizes financial and technical support. Landowners can receive financial or technical assistance for biomass feedstock development based on the renewable biomass definition for that specific program.

---

[5] CRS Report R41106, *Meeting the Renewable Fuel Standard (RFS) Mandate for Cellulosic Biofuels: Questions and Answers*, by Kelsi Bracmort.

[6] CRS Report R40455, *Renewable Energy and Energy Efficiency Tax Incentive Resources*, by Lynn J. Cunningham and Beth A. Roberts.

One program that was thought to have the potential to increase the level of private landowner participation in the biomass feedstock market was the Biomass Crop Assistance Program (BCAP), established in §9011 of the 2008 farm bill.[7] BCAP, administered by the USDA Farm Service Agency, is intended to support the establishment and production of eligible crops for conversion to bioenergy. (Crops or lands that receive payments under Title I of the 2008 farm bill are not eligible for participation in BCAP.) However, BCAP is set to expire at the end of 2012 and lacks baseline funding going forward. The fate of this program depends on the reauthorization of the farm bill. The 2008 farm bill contains other renewable energy provisions that may stimulate biomass feedstock efforts, although many of these provisions have not yet been implemented.[8]

Proposed Modification of the Biomass Definition

The definition for biomass contained in legislation determines what sources of material are deemed eligible as biomass and which lands are eligible for biomass removal. Biomass definitions typically contain three components: agriculture (e.g., crops), forestry (e.g., slash, pre-commercial thinnings), and waste (e.g., food, yard). Multiple biomass definitions can be included in a single piece of legislation to meet the requirements of associated programs or provisions. Environmental groups, private entities aspiring to participate in biomass-to-energy initiatives, and federal agencies that administer biomass-to-energy programs are likely to closely monitor the biomass definitions proposed during the farm bill and energy debates in Congress.

Significant attention was focused on the proposed biomass definitions contained in multiple legislative proposals put forth by the 111ᵗʰ Congress, including the American Clean Energy and Security Act of 2009 (ACES; H.R. 2454), the American Clean Energy Leadership Act of 2009 (ACELA; S. 1462), the Clean Energy Jobs and American Power Act (S. 1733), and the American Power Act (discussion draft). See **Table 2**.

Discussion regarding the definition of biomass during the 111ᵗʰ Congress tended to center on the type of forestry products considered as an eligible biomass source and the lands (e.g., federal, forested) where biomass removal can occur. The eligibility of forest products may have been a contentious aspect of the biomass definition primarily because of differing viewpoints on the sustainability of woody biomass supplies. Some voice disapproval about forest lands being eligible for biomass removal generally because it is uncertain whether forestry products can be removed and transported to an energy conversion facility with minimal environmental impact, and whether such removals damage forest health. Others contend that inclusion of biomass removal from federal and forested lands is necessary to meet specific biofuel mandates established in the RFS.

Debates about the definition of biomass have not been as extensive in the 112ᵗʰ Congress as they were in previous Congresses. Bills introduced during the 112ᵗʰ Congress that propose to modify the biomass definition include S. 559, S. 781, H.R. 1861, and H.R. 1920. Forthcoming discussions about energy, particularly legislation involving the Renewable Fuel Standard or energy tax incentives, may prompt further discussion about the biomass definition.

---

[7] For more information on BCAP, see CRS Report R41296, *Biomass Crop Assistance Program (BCAP): Status and Issues*, by Randy Schnepf.

[8] For more information, see CRS Report RL34130, *Renewable Energy Programs in the 2008 Farm Bill*, by Randy Schnepf; and CRS Report R42552, *The 2012 Farm Bill: A Comparison of Senate-Passed S. 3240 and the House Agriculture Committee's H.R. 6083 with Current Law*, coordinated by Ralph M. Chite.

Table 1. Biomass Definitions Contained in Legislation Enacted Since 2004

No.	Public Law/ Tax Code	Definition	Comments
1	P.L. 110-246 Food, Conservation, and Energy Act of 2008 Title IX Sec. 9001(12)	The term *renewable biomass* means— `(A) materials, pre-commercial thinnings, or invasive species from National Forest System land and public lands (as defined in section 103 of the Federal Land Policy and Management Act of 1976 (43 U.S.C. 1702)) that—`(i) are byproducts of preventive treatments that are removed—`(I) to reduce hazardous fuels; `(II) to reduce or contain disease or insect infestation; or `(III) to restore ecosystem health; `(ii) would not otherwise be used for higher-value products; and `(iii) are harvested in accordance with—`(I) applicable law and land management plans; and `(II) the requirements for—`(aa) old-growth maintenance, restoration, and management direction of paragraphs (2), (3), and (4) of subsection (e) of section 102 of the Healthy Forests Restoration Act of 2003 (16 U.S.C. 6512); and `(bb) large-tree retention of subsection (f) of that section; or `(B) any organic matter that is available on a renewable or recurring basis from non-Federal land or land belonging to an Indian or Indian tribe that is held in trust by the United States or subject to a restriction against alienation imposed by the United States, including— `(i) renewable plant material, including—`(I) feed grains; `(II) other agricultural commodities; `(III) other plants and trees; and `(IV) algae; and `(ii) waste material, including—`(I) crop residue; `(II) other vegetative waste material (including wood waste and wood residues); `(III) animal waste and byproducts (including fats, oils, greases, and manure); and `(IV) food waste and yard waste.	Definition associated with the following sections in the bill: Biorefinery Assistance Program (§9003), Repowering Assistance (§9004), Biomass Research and Development Initiative (§9008), Biomass Crop Assistance Program (§9011), Forest Biomass for Energy (§9012), and Community Wood Energy Program (§9013). In contrast to the RFS definition of *renewable biomass* (EISA, Title II, Sec. 201(1)(I)), this definition allows biomass from federal lands as a biofuel feedstock. The definition *includes* materials, pre-commercial thinnings, or invasive species from National Forest System land and public (BLM) lands; organic matter available on a renewable or recurring basis from non-federal or Indian land; renewable plant material (e.g., feed grains, other agricultural commodities, other plants and trees, and algae) and waste material (e.g., crop residue and other vegetative waste material, such as wood waste and wood residues), as well as animal waste and byproducts (including fats, oils, greases, and manure), food waste, and yard waste. This definition does *not* mention biomass from wildfire fuel treatments in the immediate vicinity of buildings, as does the RFS *renewable biomass* definition. (See definition 2.) No limits on private-sector participation evident.

No.	Public Law/ Tax Code	Definition	Comments
2	P.L. 110-140 Energy Independence and Security Act of 2007 Title II Sec. 201(1)(I)	The term **renewable biomass** means each of the following: ''(i) Planted crops and crop residue harvested from agricultural land cleared or cultivated at any time prior to the enactment of this sentence that is either actively managed or fallow, and nonforested. ''(ii) Planted trees and tree residue from actively managed tree plantations on non-federal land cleared at any time prior to enactment of this sentence, including land belonging to an Indian tribe or an Indian individual, that is held in trust by the United States or subject to a restriction against alienation imposed by the United States. ''(iii) Animal waste material and animal byproducts. ''(iv) Slash and pre-commercial thinnings that are from non-federal forestlands, including forestlands belonging to an Indian tribe or an Indian individual, that are held in trust by the United States or subject to a restriction against alienation imposed by the United States, but not forests or forestlands that are ecological communities with a global or State ranking of critically imperiled, imperiled, or rare pursuant to a State Natural Heritage Program, old growth forest, or late successional forest. ''(v) Biomass obtained from the immediate vicinity of buildings and other areas regularly occupied by people, or of public infrastructure, at risk from wildfire. ''(vi) Algae. ''(vii) Separated yard waste or food waste, including recycled cooking and trap grease.	Definition associated with this section in the bill: Renewable Fuel Standard (Title II, Subtitle A). This provision defines renewable biomass for the Renewable Fuel Standard (Title II, Subtitle A). The definition *excludes*, as biofuel feedstocks, biomass removed from federal lands and crops for forested lands (e.g., timber harvests). The definition *includes* biomass from: • slash and pre-commercial thinnings from non-federal forestlands, including Indian forestlands, but not from forestlands that are ecological communities within a global or state ranking of critically imperiled, imperiled, or rare pursuant to a State Natural Heritage Program, and not from old-growth or late successional forests; • planted trees and tree residue from actively managed tree plantations on non-federal land; • biomass obtained from the immediate vicinity of buildings, public infrastructure, and areas regularly occupied by people that are at risk from wildfire (e.g., from wildfire fuel reduction activities on non-federal lands); and • other activities, including planted crops and crop residue from nonforested agricultural land that is either actively managed or fallow; animal waste material and byproducts; separated yard waste or food waste (including recycled cooking and trap grease); and algae. Limits on private-sector participation evident—biomass may not be removed from federal lands. Some biomass material may come from non-federal forest lands.

No.	Public Law/ Tax Code	Definition	Comments
3	P.L. 110-140 Energy Independence and Security Act of 2007 Title XII Sec. 1201	The term ***biomass***—'(aa) means any organic material that is available on a renewable or recurring basis, including—'(AA) agricultural crops; '(BB) trees grown for energy production; '(CC) wood waste and wood residues; '(DD) plants (including aquatic plants and grasses); '(EE) residues; '(FF) fibers; '(GG) animal wastes and other waste materials; and '(HH) fats, oils, and greases (including recycled fats, oils, and greases); and '(bb) does not include—'(AA) paper that is commonly recycled; or '(BB) unsegregated solid waste.	Definition associated with this section in the bill: Express Loans for Renewable Energy and Energy Efficiency (§1201). The definition *excludes* paper that is commonly recycled and unsegregated solid waste. The definition *includes* any organic material available on a renewable or recurring basis, including agricultural crops, trees grown for energy production, wood waste and wood residues, plants (including aquatic plants and grasses), residues, fibers, animal wastes and other waste materials, and fats, oils, and greases (including recycled fats, oils, and greases). The definition is much less restrictive than the *renewable biomass* definition listed under EISA, Title II, Sec. 201(1)(I) for the RFS. The definition is similar to the biomass definition under the EPAct05 Renewable Energy Security Provision (§206). Does not include paper that is commonly recycled or unsegregated solid waste. Does not specify "actively managed" crops and trees as a criterion as mentioned in definition 2.
4	P.L. 110-140 Energy Independence and Security Act of 2007 Title XII Sec. 1203(e)(z)(4)(A)	The term ***biomass***—'(i) means any organic material that is available on a renewable or recurring basis, including—'(I) agricultural crops; '(II) trees grown for energy production; '(III) wood waste and wood residues; '(IV) plants (including aquatic plants and grasses); '(V) residues; '(VI) fibers; '(VII) animal wastes and other waste materials; and '(VIII) fats, oils, and greases (including recycled fats, oils, and greases); and '(ii) does not include—'(I) paper that is commonly recycled; or '(II) unsegregated solid waste.	Definition associated with this section in the bill: Small Business Energy Efficiency Program (§1203). Applicable biomass is identical to materials described in definition 3.

No.	Public Law/ Tax Code	Definition	Comments
5	Tax Code 2007 Title 26 Subtitle A Chapter I Subchapter A Part IV Subpart D Sec. 45(c)(2)	The term *closed-loop biomass* means any organic material from a plant which is planted exclusively for purposes of being used at a qualified facility to produce electricity.	Definition associated with this section in the code: Electricity Produced from Certain Renewable Resources (Sec. 45). Definition associated with this tax credit: Renewable Electricity, Refined Coal, and Indian Coal Production Credit (IRS Form 8835). Denotes the following as applicable biomass: • any organic material from a plant that is grown exclusively to produce electricity.
6	Tax Code 2007 Title 26 Subtitle A Chapter I Subchapter A Part IV Subpart D Sec. 45(c)(3)	The term *open-loop biomass* means—(i) any agricultural livestock waste nutrients, or (ii) any solid, nonhazardous, cellulosic waste material or any lignin material which is segregated from other waste materials and which is derived from—(I) any of the following forest-related resources: mill and harvesting residues, precommercial thinnings, slash, and brush, (II) solid wood waste materials, including waste pallets, crates, dunnage, manufacturing and construction wood wastes (other than pressure-treated, chemically-treated, or painted wood wastes), and landscape or right-of-way tree trimmings, but not including municipal solid waste, gas derived from the biodegradation of solid waste, or paper which is commonly recycled, or (III) agriculture sources, including orchard tree crops, vineyard, grain, legumes, sugar, and other crop by-products or residues. Such term shall not include closed-loop biomass or biomass burned in conjunction with fossil fuel (cofiring) beyond such fossil fuel required for startup and flame stabilization.	Definition associated with this section in the code: Electricity Produced from Certain Renewable Resources (Sec. 45). Definition associated with this tax credit: Renewable Electricity, Refined Coal, and Indian Coal Production Credit (IRS Form 8835). Denotes the following as applicable biomass: • any agricultural livestock waste nutrients. • any solid, nonhazardous, cellulosic waste material or lignin material. Does not include municipal solid waste, gas derived from the biodegradation of solid waste, or paper which is commonly recycled. Does not include closed-loop biomass or biomass burned in conjunction with fossil fuel (cofiring) beyond such fossil fuel required for startup and flame stabilization.

No.	Public Law/ Tax Code	Definition	Comments
7	Tax Code 2007 Title 26 Subtitle A Chapter I Subchapter A Part IV Subpart D Sec. 45k(c)(3)	The term **biomass** means any organic material other than—(A) oil and natural gas (or any product thereof), and (B) coal (including lignite) or any product thereof.	Definition associated with this section in the code: Tax Credit for Producing Fuel from a Nonconventional Source (Sec. 45k). Denotes the following as applicable biomass: • any organic material other than oil and natural gas, and coal or any product thereof. Definition does not distinguish between open-loop biomass and closed-loop biomass. Definition appears to be more expansive than definitions provided in P.L. 110-140 and P.L. 110-246.
8	Tax Code 2007 Title 26 Subtitle A Chapter I Subchapter A Part IV Subpart E Sec. 48b(c)(4)	The term **biomass** means any—(i) agricultural or plant waste, (ii) byproduct of wood or paper mill operations, including lignin in spent pulping liquors, and (iii) other products of forestry maintenance. (B) Exclusion: The term "biomass" does not include paper which is commonly recycled.	Definition associated with this section in the code: Qualifying Gasification Project Credit (Sec. 48b). Denotes the following as applicable biomass: • any agricultural or plant waste. • wood or paper mill operations byproduct. • other products of forestry maintenance. Does not include paper which is commonly recycled. Does not include closed-loop biomass.

Biomass: Comparison of Definitions in Legislation Through the 112ᵗʰ Congress

No.	Public Law/ Tax Code	Definition	Comments
9	P.L. 109-58 Energy Policy Act of 2005 Title II Sec. 203(b)(1)	The term *biomass* means any lignin waste material that is segregated from other waste materials and is determined to be nonhazardous by the Administrator of the Environmental Protection Agency and any solid, nonhazardous, cellulosic material that is derived from—(A) any of the following forest-related resources: mill residues, precommercial thinnings, slash, and brush, or nonmerchantable material; (B) solid wood waste materials, including waste pallets, crates, dunnage, manufacturing and construction wood wastes (other than pressure-treated, chemically treated, or painted wood wastes), and landscape or right-of-way tree trimmings, but not including municipal solid waste (garbage), gas derived from the biodegradation of solid waste, or paper that is commonly recycled; (C) agriculture wastes, including orchard tree crops, vineyard, grain, legumes, sugar, and other crop by-products or residues, and livestock waste nutrients; or (D) a plant that is grown exclusively as a fuel for the production of electricity.	Definition associated with this section in the bill: Federal Government Purchase Requirement for Renewable Energy (§203). Denotes the following as applicable biomass: • any lignin waste material that is segregated from other waste materials. • any solid, nonhazardous, cellulosic material derived from forest-related resources, solid wood waste materials, agriculture wastes, or a plant that is grown exclusively as a fuel for the production of electricity. Does not include municipal solid waste, gas derived from the biodegradation of solid waste, or paper that is commonly recycled. Introduces a concept that will be defined in 2007 as "closed-loop biomass" in the Internal Revenue Code. Does not specify "actively managed" crops and trees as a criterion as mentioned in definition 2. Does not discuss biomass obtained from the immediate vicinity of buildings (see definition 2). Limits on private-sector participation not specified.
10	P.L. 109-58 Energy Policy Act of 2005 Title II Sec. 206(a)(6)(B)	The term *biomass* means any organic matter that is available on a renewable or recurring basis, including agricultural crops and trees, wood and wood wastes and residues, plants (including aquatic plants), grasses, residues, fibers, and animal wastes, municipal wastes, and other waste materials.	Definition associated with this section in the bill: Renewable Energy Security Provision (§206). Denotes the following as applicable biomass: • any organic matter available on a renewable or recurring basis including agricultural crops and trees, wood and wood wastes and residues, plants (including aquatic plants), grasses, residues, fibers, and animal wastes, municipal wastes, and other waste materials. Limits on private-sector participation not specified.

No.	Public Law/ Tax Code	Definition	Comments
11	P.L. 109-58 Energy Policy Act of 2005 Title II Sec. 210(a)(1)	The term **biomass** means nonmerchantable materials or precommercial thinnings, such as trees, wood, brush, thinnings, chips, and slash, that are removed—(A) to reduce hazardous fuels; (B) to reduce or contain disease or insect infestation; or (C) to restore forest health.	Definition associated with this section in the bill: Grants to Improve Commercial Value of Forest Biomass for Electric Energy, Useful Heat, Transportation Fuels and Other Commercial Purposes Program (§210). Denotes the following as applicable biomass: • unmarketable materials or precommercial thinnings that are byproducts of preventive treatments, such as trees, wood, brush, thinnings, chips, and slash. Definition limited to forestry biomass sources. Limits on private-sector participation not specified.
12	P.L. 109-58 Energy Policy Act of 2005 Title IX Subtitle C Sec. 932(a)(1)	The term **biomass** means—(A) any organic material grown for the purpose of being converted to energy; (B) any organic byproduct of agriculture (including wastes from food production and processing) that can be converted into energy; or (C) any waste material that can be converted to energy, is segregated from other waste materials, and is derived from—(i) any of the following forest-related resources: mill residues, precommercial thinnings, slash, brush, or otherwise nonmerchantable material; or (ii) wood waste materials, including waste pallets, crates, dunnage, manufacturing and construction wood wastes (other than pressure-treated, chemically-treated, or painted wood wastes), and landscape or right-of-way tree trimmings, but not including municipal solid waste, gas derived from the biodegradation of municipal solid waste, or paper that is commonly recycled.	Definition associated with this section in the bill: Bioenergy Program (§932). Denotes the following as applicable biomass: • any organic material grown for the purpose of being converted to energy. • any organic byproduct of agriculture. • any waste material that can be converted to energy, is segregated from other waste materials, and is derived from forest-related resources or wood waste materials, and landscape or right-of-way tree trimmings. Does not include municipal solid waste, gas derived from the biodegradation of municipal solid waste, or paper that is commonly recycled. Does not specify "actively managed" crops and trees as a criterion as mentioned in definition 2. Limits on private-sector participation not specified.

No.	Public Law/ Tax Code	Definition	Comments
13	P.L. 109-58 Energy Policy Act of 2005 Title XIII Subtitle A Sec. 1307 Sec. 48B(c)(4)	The term *biomass* means any—`(i) agricultural or plant waste, `(ii) byproduct of wood or paper mill operations, including lignin in spent pulping liquors, and `(iii) other products of forestry maintenance. `(B) EXCLUSION- The term `biomass' does not include paper which is commonly recycled.	Definition associated with this section in the bill: Credit for Investment in Clean Coal Facilities (§1307). Denotes the following as applicable biomass: • any agricultural or plant waste. • any byproduct of wood or paper mill operations. • any other products of forestry maintenance. Does not include paper which is commonly recycled. Does not specify "actively managed" crops and trees as a criterion as mentioned in definition 2. Limits on private-sector participation not specified.
14	P.L. 109-58 Energy Policy Act of 2005 Title XV Subtitle A Sec. 1512(r)(4)(B)	The term *renewable biomass* is, as defined in Presidential Executive Order 13134, published in the *Federal Register* on August 16, 1999, any organic matter that is available on a renewable or recurring basis (excluding old-growth timber), including dedicated energy crops and trees, agricultural food and feed crop residues, aquatic plants, animal wastes, wood and wood residues, paper and paper residues, and other vegetative waste materials. Old-growth timber means timber of a forest from the late successional stage of forest development.	Definition associated with this section in the bill: Conversion Assistance for Cellulosic Biomass, Waste-Derived Ethanol, Approved Renewable Fuels Grants Program (§1512). Denotes the following as applicable biomass: • any organic matter that is available on a renewable or recurring basis (excluding old-growth timber) including dedicated energy crops and trees, agricultural food and feed crop residues, aquatic plants, animal wastes, wood and wood residues, paper and paper residues, and other vegetative waste materials. Does not specify "actively managed" crops and trees as a criterion as mentioned in definition 2. Limits on private-sector participation not specified.

Table 2. Biomass Definitions Contained in Legislation in the 111th Congress

No.	Bill	Comments
1	H.R. 2454 American Clean Energy and Security Act of 2009 (ACES) Title I Subtitle A Sec. 101	Definition associated with this section in the bill: Combined Efficiency and Renewable Electricity Standard (§610). The definition is an amalgamation of previous language, primarily from the 2008 farm bill. The forestry component of the definition includes materials, pre-commercial thinnings, or invasive species from National Forest System land and public (BLM) lands, including: • the byproducts of preventive treatments (such as trees, wood, brush, thinnings, chips, and slash), • those removed as part of a federally recognized timber sale, or • those removed to reduce hazardous fuels, to reduce or contain disease or insect infestation, or to restore ecosystem health. The materials, pre-commercial thinnings, or invasive species must be harvested in environmentally sustainable quantities, as determined by the federal land manager; and in accordance with federal and state law, and applicable land management plans. The definition *excludes* lands in the National Wilderness Preservation System, Wilderness Study Areas, Inventoried Roadless Areas, old-growth stands, late-successional stands (except for dead, severely damaged, or badly infested trees), components of the National Landscape Conservation System, National Monuments, National Conservation Areas, Designated Primitive Areas, or Wild and Scenic Rivers corridors. These ineligible lands were originally in the Healthy Forests Restoration Act of 2003 (P.L. 108-148). As passed, the definition includes "federally recognized timber sale," "harvested in environmentally sustainable quantities," and "residues and byproducts from wood, pulp, or paper products facilities." As passed, the agricultural component of the definition includes: organic matter available on a renewable or recurring basis from non-federal or Indian land, including renewable plant material (e.g., feed grains, other agricultural commodities, other plants and trees, and algae) and waste material (e.g., crop residue and other vegetative waste material, such as wood waste and wood residues), animal waste and byproducts (including fats, oils, greases, and manure), construction waste, and food waste and yard waste. The definition *includes* residues and byproducts from wood, pulp, or paper products facilities. Advocates for this biomass definition include groups who seek to use the potentially substantial volumes of waste woody biomass from federal lands and other (non-plantation) forested lands (e.g., waste from timber harvests, from pre-commercial thinnings, or from wildfire fuel reduction treatments) as a source of renewable energy. Opponents include groups who seek to preserve undisturbed forested land and federal land, and who are concerned that incentives for using wood waste would encourage activities that damage forest lands, possibly harming important wildlife habitats and water quality.
2	H.R. 2454 American Clean Energy and Security Act of 2009 Title I Subtitle C Sec. 126	Definition associated with this section in the bill: Clean Air Act (42 U.S.C. §7545(o)(1)) Renewable Fuel Standard. The majority of the definition is identical to the text of the *renewable biomass* definition listed under ACES, Title I, Subtitle A, Sec. 101 (see definition 1 above). The sole exception is the addition of clause (ff) that *includes* as applicable biomass "the non-fossil biogenic portion of municipal solid waste and construction, demolition, and disaster debris." Applicability and restrictions for this definition are identical to those described for the *renewable biomass* definition under ACES, Title I, Subtitle A, Sec. 101 (see definition 1 above). This definition would replace the EISA current definition of renewable biomass under the RFS.

No.	Bill	Comments
3	**H.R. 2454** American Clean Energy and Security Act of 2009 Title VIII Part E Sec. 751	Definition associated with this section in the bill: Supplemental Emissions Reductions from Reduced Deforestation provision. The definition, applicability, and restrictions are identical to those described for renewable biomass definition listed under ACES, Title I, Subtitle A, Sec. 101 (see definition 1 above).
4	**S. 1462** American Clean Energy Leadership Act of 2009 (ACELA) Title I Subtitle C Sec. 133	Definition associated with this section in the bill: federal purchase requirement for renewable electricity provision (Energy Policy Act of 2005 §203; 42 U.S.C. 15852). The forestry component of the definition includes nonhazardous organic materials: • residues and byproducts from milled logs; • wood, paper products that are not commonly recyclable, and vegetation diverted from or separated from other waste in the municipal waste stream; • hazard trees, trimmings, and brush that are necessary to remove to maintain a utility right-of-way or a public road; • trees, trimmings, and brush harvested from the immediate vicinity of any building, campground, or other structure in wildfire-prone areas to reduce the risk to the structure or campground or to human life from wildfires; • invasive species removed to control or eradicate the invasive species; • slash, brush, trees, and other vegetation that is harvested from non-federal or Indian land that is, at the time of harvest: —naturally regenerated forest land; forest land planted to restore a naturally regenerated forest; or if harvested in quantities and through practices that maintain or contribute to the restoration of the species, ecological systems, and ecological communities for which the conservation forest land was identified, conservation forest land (as defined in the proposed amendment); or —planted forest land; and cropland (including fallow land), pastureland, or planted forest land. Federal land requirements for this definition also include that the harvest be in accordance with applicable law and land management plans and be done in quantities and through practices that maintain or contribute to the restoration of ecological sustainability. The harvest can include (i) slash; and (ii) brush and trees that are byproducts of ecological restoration, disease or insect infestation control, or hazardous fuels reduction treatments. The harvest must not exceed the minimum size standards for sawtimber and must be from stands killed by an insect or disease epidemic or a natural disaster and that do not meet the utilization standards for sawtimber. The agricultural component includes other nonhazardous organic materials: crops, crop byproducts, and crop residues from non-forested non-federal or Indian land that is cropland (including fallow land) or pastureland. The definition also includes: animal waste and animal byproducts; food waste; and algae. The term *nonhazardous* is included in the biomass definition in the EPAct05 and in the IRS tax code definition for open loop biomass (Title 26 Subtitle A Chapter I Subchapter A Part IV Subpart D Sec. 45(c)(3)).

No.	Bill	Comments
		Advocates for this definition include groups who seek to use the potentially substantial volumes of waste woody biomass from federal lands and other forested lands as a source of renewable energy. Opponents include groups who oppose disturbance of forested land and federal land, and are concerned that incentives for using wood waste encourage activities that disturb forest lands, possibly damaging important wildlife habitats and water quality. While the definition may be narrower and more limiting in some ways than the definitions in ACES and the 2008 farm bill, it also may provide more possibilities for accessing federal and other forest lands for biomass energy production.
5	S. 1733 Title VII Sec. 700	Definition associated with this section in the bill: Offsets (S. 1733, Title VII, Part D).
		The forestry component of the definition includes:
		• Trees, brush, slash, residues, or any other vegetative matter removed from within 600 feet of any building, campground, or route designated for evacuation by a public official with responsibility for emergency preparedness, or from within 300 feet of a paved road, electric transmission line, utility tower, or water supply line.
		• Residues from or byproducts of milled logs.
		• Any of the following materials removed from forested land that is not federal and is not high conservation priority land:
		—Trees, brush, slash, residues, interplanted energy crops, or any other vegetative matter removed from an actively managed tree plantation established prior to January 1, 2009; or on land that, as of January 1, 2009, was cultivated or fallow and non-forested.
		—Trees, logging residue, thinnings, cull trees, pulpwood, and brush removed from naturally regenerated forests or other non-plantation forests, including for the purposes of hazardous fuel reduction or preventative treatment for reducing or containing insect or disease infestation.
		—Logging residue, thinnings, cull trees, pulpwood, brush, and species that are non-native and noxious, from stands that were planted and managed after January 1, 2009, to restore or maintain native forest types.
		—Dead or severely damaged trees removed within 5 years of fire, blowdown, or other natural disaster, and badly infested trees.
		• Materials, pre-commercial thinnings, or removed invasive species from National Forest System land and public lands (as defined in section 103 of the Federal Land Policy and Management Act of 1976 (43 U.S.C. 1702)), including those that are byproducts of preventive treatments (such as trees, wood, brush, thinnings, chips, and slash), that are removed as part of a federally recognized timber sale, or that are removed to reduce hazardous fuels, to reduce or contain disease or insect infestation, or to restore ecosystem health, and that are not from components of the National Wilderness Preservation System, Wilderness Study Areas, Inventoried Roadless Areas, old growth or mature forest stands, components of the National Landscape Conservation System, National Monuments, National Conservation Areas, Designated Primitive Areas; or Wild and Scenic Rivers corridors; harvested in environmentally sustainable quantities, as determined by the appropriate Federal land manager; and are harvested in accordance with Federal and State law, and applicable land management plans.
		The definition excludes lands in the National Wilderness Preservation System, Wilderness Study Areas, Inventoried Roadless Areas, old growth or mature forest stands, components of the National Landscape Conservation System, National Monuments, National Conservation Areas, Designated Primitive Areas; or Wild and Scenic Rivers corridors. These ineligible lands were originally in the Healthy Forests Restoration Act of 2003 (P.L. 108-148). The definition includes "harvested in environmentally sustainable quantities."

No.	Bill	Comments
		The agricultural component includes plant material including waste material, harvested or collected from actively managed agricultural land that was in cultivation, cleared, or fallow and nonforested on January 1, 2009; plant material including waste material, harvested or collected from pastureland that was nonforested on January 1, 2009; Nonhazardous vegetative matter derived from waste, including separated yardwaste, landscape right-of-way trimmings, construction and demolition debris, or food waste (but not municipal solid waste, recyclable wastepaper, painted, treated or pressurized wood, or wood contaminated with plastic or metals). The definition also includes: animal waste and animal byproducts (including products of animal waste digesters); and algae.
		The renewable biomass definition contained in S. 1733 differs from other proposed definitions in that it introduces an eligibility date (January 1, 2009). The inclusion of a date may preclude obtaining financial or technical assistance, or offset credit if biomass was planted, harvested, collected, removed, managed, or cultivated prior to the eligibility date. The forestry component of the renewable biomass definition listed in S. 1733 is identical to the forestry component of the definition contained in ACES. Advocates for the biomass definition contained in S. 1733 include groups pursuing the use of large amounts of waste woody biomass from federal lands and other forested lands as a source of renewable energy. Opponents include groups who are opposed for environmental reasons to disturbing federal and forested land. Other opponents include groups that seek to use plant or woody materials established prior to January 1, 2009, for energy purposes.
6	American Power Act (Discussion Draft) Title VII Sec. 700	Definition associated with this section in the bill: Offsets (Title VII, Part E). The forestry component of the definition is nearly identical to the forestry component listed under ACES, Title I, Subtitle A, Sec. 101 (see definition 1 above), with the following two exceptions. • Land management plans define an old growth stand and late-successional stands. The definition excludes lands in the National Wilderness Preservation System, Wilderness Study Areas, Inventoried Roadless Areas, old-growth stands (as defined by the applicable land management plan), late-successional stands, except for dead, severely damaged, or badly infested trees (as defined by the applicable land management plan), components of the National Landscape Conservation System, National Monuments, National Conservation Areas, Designated Primitive Areas, or Wild and Scenic Rivers corridors. These ineligible lands were originally in the Healthy Forests Restoration Act of 2003 (P.L. 108-148). • The definition does not specifically identify harvesting in accordance with federal and state law. The materials, pre-commercial thinnings, or invasive species must be harvested in environmentally sustainable quantities, as determined by the federal land manager; and in accordance with applicable law and land management plans. The agricultural component of the renewable biomass definition is identical to the agriculture component listed under ACES, Title I, Subtitle A, Sec. 101 (see definition 1 above).
7	American Power Act (Discussion Draft) Title VII Sec. 2214	Excess biomass definition associated with this section in the bill: Enhanced soil sequestration (Title VII, Subtitle C, Part II). Excess biomass is any plant matter targeted for removal from public land to promote ecosystem health including trees or tree waste on public land; wood and wood wastes and residues; and weedy plants and grasses (including aquatic, noxious, or invasive plants). The American Power Act (discussion draft) definition for renewable biomass is similar to the definition contained in H.R. 2454, with two exceptions. The first is the use of land management plans to define both an old growth stand and a late-successional stand. The second exception is that the forestry component of the definition does not specifically identify harvesting in accordance with federal and state law. The American Power Act discussion draft introduces a second biomass term—excess biomass—for the enhanced soil sequestration provision in the bill. Excess biomass may contain material removed from public land to promote ecosystem health.

Author Contact Information

Kelsi Bracmort
Specialist in Agricultural Conservation and Natural
Resources Policy
kbracmort@crs.loc.gov, 7-7283

Acknowledgments

Ross Gorte, retired CRS Specialist in Natural Resources Policy, made important contributions to this report.

www.ingramcontent.com/pod-product-compliance
Lightning Source LLC
Chambersburg PA
CBHW081307170526
45165CB00010B/3292